TOOLS FOR CAREGIVERS

- **ATOS:** 0.9
- **GRL:** C
- **WORD COUNT:** 30

- **CURRICULUM CONNECTIONS:** insects, nature, patterns

Skills to Teach

- **HIGH-FREQUENCY WORDS:** a, does, I, into, one, see, this, what
- **CONTENT WORDS:** becomes, bugs, caterpillars, hair, moth, spikes, spots, stripes, turn, wow
- **PUNCTUATION:** exclamation points, periods, question mark
- **WORD STUDY:** long /a/, spelled *ai* (*hair*); long /e/, spelled *ee* (*see*); multisyllable word (*caterpillars*)
- **TEXT TYPE:** information report

Before Reading Activities

- Read the title and give a simple statement of the main idea.
- Have students "walk" though the book and talk about what they see in the pictures.
- Introduce new vocabulary by having students predict the first letter and locate the word in the text.
- Discuss any unfamiliar concepts that are in the text.

After Reading Activities

Caterpillars come in many colors and patterns. What patterns were mentioned in the book? Have the readers look around them. Can they find these patterns on anything else they see? What other patterns can they distinguish in their setting?

Tadpole Books are published by Jump!, 5357 Penn Avenue South, Minneapolis, MN 55419, www.jumplibrary.com

Copyright ©2020 Jump. International copyright reserved in all countries. No part of this book may be reproduced in any form without written permission from the publisher.

Editor: Jenna Trnka **Designer:** Michelle Sonnek

Photo Credits: chrom/Shutterstock, cover; Matee Nuserm/Shutterstock, 1; Jay Ondreicka/Shutterstock, 2tl, 3, 14–15 (left); Re Metau/Shutterstock, 2br, 4–5; Eric Isselee/Shutterstock, 2bl, 6–7; Cathy Keifer/Shutterstock, 2mr, 8–9, 14–15 (right); Anest/Shutterstock, 2tr, 10–11; Captainflash/iStock, 12; Nick Hawkins/Minden Pictures, 2ml, 13; TessarTheTegu/Shutterstock, 16.

Library of Congress Cataloging-in-Publication Data
Names: Nilsen, Genevieve, author.
Title: I see caterpillars / by Genevieve Nilsen.
Description: Tadpole books edition. | Minneapolis, MN: Jump!, Inc., (2020) | Series: Backyard bugs | Audience: Age 3–6. | Includes index.
Identifiers: LCCN 2018050519 (print) | LCCN 2018051536 (ebook) | ISBN 9781641287975 (ebook) | ISBN 9781641287951 (hardcover: alk. paper) | ISBN 9781641287968 (paperback)
Subjects: LCSH: Caterpillars—Juvenile literature.
Classification: LCC QL544.2 (ebook) | LCC QL544.2 .N56 2020 (print) | DDC 595.7813/92—dc23
LC record available at https://lccn.loc.gov/2018050519

I SEE CATERPILLARS

by Genevieve Nilsen

TABLE OF CONTENTS

tadpole
books

WORDS TO KNOW

caterpillars

hair

moth

spikes

spots

stripes

I SEE CATERPILLARS

I see bugs!

stripe

4

I see stripes.

spot

I see spots.

spike

I see spikes.

hair

I see hair.

caterpillar

This one becomes
a moth!

moth

Wow!

caterpillar

What does this one turn into?

LET'S REVIEW!

Caterpillars can be many colors. They can have spots, spikes, or hair. What do you notice about this caterpillar?

INDEX